I0493688

NIST TECHNICAL NOTE 1756

Generation of Large Directional Wind Speed Datasets from Smaller Synoptic Storm or Thunderstorm Data Samples

DongHun Yeo

National Windstorm Impact Reduction Program
Engineering Laboratory
National Institute of Standards and Technology
Gaithersburg, MD 20899-8611

August 2012

U.S. Department of Commerce
Dr. Rebecca Blank, *Acting Secretary*

National Institute of Standards and Technology
Dr. Patrick D. Gallagher, *Director*

Disclaimers

(1) The policy of the NIST is to use the International System of Units in its technical communications. In this document however, works of authors outside NIST are cited which describe measurements in certain non-SI units. Thus, it is more practical to include the non-SI unit measurements from these references.

(2) Certain trade names or company products or procedures may be mentioned in the text to specify adequately the experimental procedure or equipment used. In no case does such identification imply recommendation or endorsement by the National Institute of Standards and Technology, nor does it imply that the products or procedures are the best available for the purpose.

Abstract

For structures sensitive to wind directionality, methods for the estimation of wind effects require the use of time series of directional wind speeds covering time periods exceeding the length of the Mean Recurrence Interval of interest in design. This study proposes a procedure for generating such time series from relatively short wind data sets. First, a procedure is developed for estimating the parameters of the distributions of the directional wind speeds, given that in the size of the data sample any one directional sector can in some instances be small. Once the distribution parameters are estimated, a simple Monte Carlo procedure is used for the data generation. The wind speed data being generated can be used within the framework of the Database-Assisted Design approach to determine wind effects on buildings by accounting for wind directionality.

Keywords: Directional wind speeds; Mean recurrence intervals; Extreme value statistics; Synthetic wind speed data; Monte Carlo simulation; Generalized Pareto distribution.

Acknowledgements

The author would like to thank E. Simiu and F. T. Lombardo of the National Institute of Standards and Technology for valuable advice and for providing wind speed datasets used in this work.

Contents

Abstract iii

Acknowledgements.. iv

Contents v

List of Figures.. vi

List of Tables vii

Nomenclature viii

1. Introduction...1

2. Methodology 2
2.1 Measured Data Samples in Elemental and Consolidated Sectors............................2
2.2 Probabilistic Modeling of Extreme Wind Speeds
     Using Directional Sector Data Sets...2
2.3 Requisite Number of Generated Synthetic Windstorms 4
2.4 Generation of Large Directional Data Sets by Monte Carlo Simulations 5

3. Application...7
3.1 Small Sets of Measured Directional Wind Speed Data 7
3.2 GPD Parameter Estimation 9
3.3 Generation of Large Directional Wind Speed Datasets
     by Monte Carlo Simulation..12

4. Conclusions..15

References 16

List of Figures

Fig. 1 Construction of non-directional wind speed datasets from directional data 4

Fig. 2 Procedure for estimation of GPD parameters... 6

Fig. 3 Series of synoptic wind speeds, ASOS data (Newark, NJ) 7

Fig 4 Histogram of measured wind speeds of synoptic wind data (Newark, NJ) 8

Fig. 5 Histogram of measured wind speeds in directional sectors 8

Fig. 6 Estimated tail length parameters versus threshold and number of threshold exceedances 10

Fig. 7 Estimated scale parameters versus threshold and number of threshold exceedances 10

Fig 8 Wind speeds in directional sectors as functions of MRIs (for original
c: ———— for adjusted c: ············· for $c = -0.01$) 11

Fig. 9 Comparison of λ_i's in elemental sectors for measured data and synthetic data 13

Fig. 10 Histogram of synthetic wind speed data for MRI = 10,000 years (Newark, NJ) 13

Fig. 11 Synthetic wind speeds in elemental sectors 14

List of Tables

Table 1. Estimated parameters of generalized Pareto distributions 9

Nomenclature

ASOS	Automated Surface Observing Systems
F_i	Cumulative Density Function of GPD for sector i
GPD	Generalized Pareto distribution
MRI	Mean Recurrence Interval
$\bar{N}$	MRI of wind speeds
$\bar{N}_i$	MRI of wind speeds for sector i
POT	Peaks over Threshold
a_i	GPD parameter for scale for sector i
c_i	GPD parameter for tail length (or shape) for sector i
c_{ij}	Tail length parameter with threshold u_{ij} for sector i
d	Total number of elemental sectors
m	Total number of directional sectors
m_i	Total number of threshold wind speeds for estimation of GPD parameters for sector i
mph	Miles per hour
$n_{d,i}$	Number of wind speeds above U in directional sector i
n_{0k}	Number of wind speeds not higher than U in elemental sector k
n_k	Number of wind speeds above U in elemental sector k
q_i	Probability that wind speeds in sector i are not larger than U
r	Number of synthetic directional windstorms
r_{meas}	Number of measured directional windstorms
s	Multiplier to obtain a sufficient number of synthetic windstorms
u_i	Threshold of wind speeds for sector i
u_{ij}	Threshold j of wind speeds for sector i
$u_{max,i}$	Maximum threshold of wind speeds for constant c_i for sector i
$u_{min,i}$	Minimum threshold of wind speeds for constant c_i for sector i

U	Maximum value of threshold wind speeds from all sectors
v_i	Time series of directional wind speeds for sector i
λ	Mean yearly arrival rate of windstorm events
λ_i	Mean yearly arrival rate of windstorm events for sector i
μ	Random noise

1. Introduction

Inherent in the ASCE 7-10 Standard (ASCE 2010) wind loading provisions are approximations that may be acceptable for the design of ordinary structures but are deemed unacceptable for the design of special structures, including tall buildings. Among these approximations are those inherent in the use by ASCE 7-10 Standard of a blanket wind directionality factor in conjunction with non-directional wind speeds with a 300-year, 700-year, or 1,700-year Mean Recurrence Interval (MRI). Such use results in MRIs of wind effects equal to the MRIs of the design wind speeds that nominally induce those effects. For structures for which wind directionality effects are determined by taking into account explicitly directional aerodynamic, wind climatological, and dynamic information, the nominal identity between the MRI of the design wind speeds and the MRI of the wind effects of interest no longer holds. An effective method for doing so requires the development of time series of directional wind speeds covering time intervals longer than the MRIs considered in design (see, e.g., Yeo and Simiu, 2011). This is achieved for regions not prone to hurricanes through the use of simulation techniques that employ relatively short records of measured directional wind speeds. To avoid estimation errors due to the commingling of wind speed data associated with different types of storm it is typically necessary to perform separate simulations for synoptic storm and thunderstorm data (Lombardo, Main and Simiu 2009). If the correlation between any two directional data sets is less than, say, 0.7, which is typically the case for sufficiently large central angles of the directional sectors, the effect of the correlation on the estimated wind effects is negligible, and the wind speed data contained in any two sectors may be assumed to be independent; this assumption is marginally conservative from a structural engineering viewpoint (Grigoriu 2009), and is used throughout this work. If the correlations are relatively large, which may be the case if the central angles are small (e.g., 5° or 10°), a methodology may be used that accounts for the correlations and is based on a translation model for the d-dimensional wind speed vector whose components are the directional wind speeds in a windstorm event (d is the total number of elemental sectors); for details see Grigoriu (2009).

The methodology used in this work involves the following steps. First, a decision is made on the number of *elemental sectors* being considered. For example, it is in practice reasonable to consider elemental sectors with central angle $360/18 = 20^{\circ}$. Second, it is checked whether the number of data in each elemental sector is sufficiently large to allow reasonably precise estimates of the requisite probability distributions. Third, if for some elemental sectors this is not the case, *consolidated sectors* are created as the union of two or more consecutive elemental sectors, so that the number of data in the consolidated sectors be sufficiently large. Fourth, for each sector, whether consolidated or elemental, an Extreme Value probabilistic model is fitted to the respective data sample. Since ASOS data typically exceed a threshold of approximately 35 mph, it is appropriate to use the generalized Pareto distribution (GPD) in conjunction with the Peaks over Threshold (POT) approach. Fifth, each of the probabilistic models so obtained is used to generate the requisite simulated data sets corresponding to the respective consolidated or elemental data set. The size r of the time series of simulated storm events associated with the simulated data must be such that the ratio r/λ is equal to or exceeds s times the MRI of the wind effects considered in design, where $s \geq 2$, say, and λ is the mean yearly arrival rate of the storm events at the location of interest.

1

The following sections describe the methodology in some detail, and include an example of its application. The report ends with a set of conclusions. The software for implementing the methodology is available on www.nist.gov/wind.

2. Methodology

2.1 *Measured Data Samples in Elemental and Consolidated Sectors*

The first step of the methodology consists of inspecting the set of measured data on the basis of which the simulations yielding the requisite synthetic sets are to be performed. That set is extracted from recorded measurements obtained over periods of, say, 10 years to 50 years. In the United States measured data are currently typically obtained from *Automated Surface Observing Systems* (ASOS) by using software described by Lombardo, Main and Simiu (2009) and available on www.nist.gov/wind. The data are subjected to post-processing aimed at ensuring micrometeorological homogeneity with respect to surface roughness, averaging time, and height above the surface. In addition, if the post-processed data are reported in terms of integer values, noise contained in the interval -0.5 mph $< \mu \leq 0.5$ mph is added to them to allow their distinct ordering. For example, if a data set contains three 20 mph wind speeds, the addition of noise would result in three speeds with the distinct values 19.8764, 20.0004, and 19.1087, say.

Once the measured data set is available in its final form it is necessary to determine whether the data sets corresponding to each of the individual directional sectors of interest are sufficiently large to allow the application of POT methods of extreme value statistical analysis. If an elemental sector contains a sufficiently large number of data, then those data will be subjected to the requisite statistical analysis. If the number of data is not sufficient, then the data set considered for analysis consists of the data contained in a consolidated sector. A *directional sector* denotes either an elemental or a consolidated sector containing a number of data that is sufficiently large to allow a useful statistical analysis. For a consolidated sector consisting of two or more elemental sectors, only the largest of the two or more elemental sector wind speeds associated with any given windstorm event is considered in the analysis. For example, consider a consolidated sector formed of two elemental sectors. Assume that in a given storm event the wind speed is 45 mph in one elemental sector and 49 mph in the second elemental sector. Then only the 49 mph speed is considered in the analysis of the consolidated sector data.

POT requires the definition of a threshold such that only data exceeding that threshold are considered in the extreme value analysis. The threshold being considered is determined as indicated in Section 2.2.

2.2 *Probabilistic Modeling of Extreme Wind Speeds Using Directional Sector Data Sets*

The generalized Pareto distribution (GPD) is widely assumed to be an appropriate model of the extreme wind speeds (see, e.g., Simiu, 2011 and references therein). That is, for a directional sector i (elemental or consolidated), the distribution of the extreme wind speeds v_i larger than the threshold u_i is the GPD with parameters (c_i, a_i, u_i, λ_i), where c_i is the shape (i.e., tail length) parameter, a_i is the scale parameter, u_i is the threshold, and λ_i is the mean arrival rate, estimated as the sample size divided by the duration of the data record in years.

The expression for the GPD is

$$F_i(v_i) = \begin{cases} 1 - \left[1 + c_i \left(\dfrac{v_i - u_i}{a_i} \right) \right]^{-1/c_i} & \text{for } c_i \neq 0 \\[2ex] 1 - \exp\left(\dfrac{v_i - u_i}{a_i} \right) & \text{for } c_i = 0 \end{cases} \tag{1}$$

If the mean annual rate of arrival of the windstorms is unity (corresponding to one windstorm per year), then the mean recurrence interval of the speed v_i is $\bar{N}_i = 1/[1 - F_i(v_i)]$. However, if the mean rate of arrival is λ_i, then $\bar{N}_i = 1/\{\lambda_i [1 - F_i(v_i)]\}$.

Equation 1 can be inverted to obtain the speed v_i corresponding to the probability F_i. Then,

$$v_i(\bar{N}_i) = u_i - \frac{a_i}{c_i}[1 - (\lambda_i \bar{N}_i)^{c_i}] \tag{2}$$

where $v_i \geq u_i$ for $c_i \geq 0$ (i.e., for Extreme Value Type I (Gumbel) and Type II (Fréchet) distribution tails), and $u_i \leq v_i \leq u_i - a_i/c_i$ for $c_i < 0$ (i.e., for Extreme Value Type III (reverse Weibull) distribution tails, for which v_i has an upper bound).

The parameters c_i and a_i for the directional sector i may be estimated by the de Haan method (de Haan 1994). First, a parameter c_{ij} ($j = 1, 2, 3,..., m_i$) is estimated and plotted as a function of successive threshold values u_{ij}. For example, values of the threshold in directional sector i can be $u_{ij} = 35, 36,..., 72$ mph. If u_{ij} is too high, the number of data available for analysis will be too small, meaning that the precision of the estimates will be inadequate. If u_{ij} is too low, the data set will contain data that are not representative of the extreme wind speeds, meaning that the estimates will be biased. Typically, for a certain range of threshold values the estimated tail length parameter c_{ij} will have an approximately constant value (see, e.g., Simiu, 2011, p. 155). This value is chosen to be the estimator c_i of the tail length parameter for the ith directional sector. The smallest and the largest of the thresholds u_{ij} for which the tail length parameter has the approximate value c_i are denoted by $u_{i,min}$ and $u_{i,max}$, respectively. The threshold $u_{i,min}$ determines the size of the sample used to determine the parameter a_i in Eq. (1) (see, e.g., Simiu, 2011, p. 148). As was noted in Sect. 2.1, of the wind speeds within a consolidated sector that belong to the same windstorm event, only the largest is included in data set being analyzed.

In some cases the estimated tail length parameter c_i is positive, meaning that the distribution tail is described by a Type II Extreme Value distribution. This has been shown to be typically the result of sampling errors in the parameter estimation and can result in unrealistically large estimates of extreme wind speeds with long MRIs (Simiu and Scanlan, 1996, p. 97). For this reason, if the estimated value of the GPD tail length parameter is $c_i > -0.01$, the value used in the calculations should be taken as $c_i = -0.01$ (corresponding to within a close approximation to a Gumbel distribution tail); if the estimated value is $c_i < -0.1$, the value used in the calculations should be taken as $c_i = -0.1$, thereby avoiding distribution tails that may be unconservatively short. (Note that, for practical purposes, distribution tails for which $c = -0.01$ differ insignificantly from tails of the Gumbel distribution, which correspond in the limit to $c = 0$, see Sect. 3.2.) The purpose of these recommendations, which have been adopted in this work, is to avoid strong deviations from the Gumbel distribution commonly accepted at this time in the standard- writing community in the U.S.

2.3 Requisite Number of Generated Synthetic Windstorms

The first step in obtaining the requisite number of synthetic windstorms consists of determining the value of the threshold $U = \max_i(u_{min,i})$ ($i = 1, 2, \ldots, m$), where m is the total number of directional sectors. The second step consist of creating a vector, the components of which are the largest directional wind speeds $\max_i(v_i) > U$ measured in each windstorm. An illustrative example of directional and non-directional wind speeds with $m = 4$ is shown in Fig. 1. The third step is the calculation of the mean arrival rate λ of the windstorms for which $\max_i(v_i) > U$. Finally, given the MRI $\bar{N}$ of interest in design (e.g., 700 years or 1,700 years), the requisite number of synthetic windstorms to be generated should cover a time interval equal to s times ($s \geq 2$, say) the MRI $\bar{N}$ (see Simiu, 2011, p. 158). This will assure a more precise estimation of the wind speed with MRI $\bar{N}$. Therefore, the number of windstorms to be generated is

$$r = s\lambda\bar{N} \tag{3}$$

Note that r windstorms induce precisely r sets of directional wind effects.

Directional wind speeds						Non-directional wind speeds	
Windstorm event	Sect. 1	Sect. 2	ct 3	Sec . 4		Windstorm event	Non-dir. Sect.
1	25	34	58	69	For each storm	1	69
2	31	70	21	37	$v = \max(v_i)$	2	70
3	45	44	37	24		3	45
4	60	29	52	53		4	60
5	36	51	48	41		5	51
6	45	38	33	5		6	45
7	32	63	67	41		7	67
8	40	32	25	49		8	49

Fig. 1 Construction of non-directional wind speed datasets from directional data

2.4 *Generation of Large Directional Data Sets by Monte Carlo Simulations*

The generation is performed independently for each of the elemental sectors k ($k = 1, 2,..., d$). For each k we have $r = n_{0k} + n_k$, where n_{0k} and n_k denote, respectively, the number of wind speeds in sector k that are at most equal to the threshold U and the number of wind speeds in sector k that are higher than U.

Denote by q_k the probability of measured wind speeds in sector k being larger than the threshold U, (i.e., $q_k = P(v_k > U)$). Then $n_k = q_k r = (\lambda_k/\lambda) r$, where λ_k and λ are, respectively, the mean yearly arrival rate of measured windstorms in sector k that are higher than U and the mean arrival rate of non-directional wind speeds that are higher than U. The n_k data are generated by Monte Carlo simulation using the following GPD model:

$$v_k = U - a_k[1-(1-F)^{-c_k}]/c_k \qquad (4)$$

where F is a random number uniformly distributed between 0 and 1. The positioning of the n_k speeds among the r speeds of sector k is randomly selected between 1 and r. For convenience, the n_{0k} wind speeds at most equal to U are assigned zero speeds.

The GPD parameters a_k and c_k used in Eq. (4) are estimated by the following procedure (Fig. 2). If, for an elemental sector k, we have $u_{max,k} \geq U \equiv \max_i(u_{min,i})$, then the parameters of the GPD model are based on U. If $u_{max,k} < U$, then two cases need to be considered. In the first case $c_k(u_{max,k}) \geq c_k(U)$. In this case the use of the parameter $c_k(U)$ would be unconservative. For this reason, even though the threshold is specified as U and the mean arrival rate of the wind speeds in excess of that threshold is based on that threshold, the tail length and the scale parameter will be based on $u_{max,k}$. In the second case $c_k(u_{max,k}) < c_k(U)$. Then, for the sake of conservatism, it is assumed that $c_k(U) = -0.01$, which is the largest possible value considered in the extreme value statistical analysis (see Sect. 2.2), and that the scale parameter is also chosen to correspond to $u_{max,k}$. This choice is justified by the observation that basing the scale parameter on U can in this case result in unreliable estimates.

In the final step, wind speeds larger than U need to be randomly relocated in their respective elemental sectors if any windstorm event has zero wind speeds in all elemental sectors.

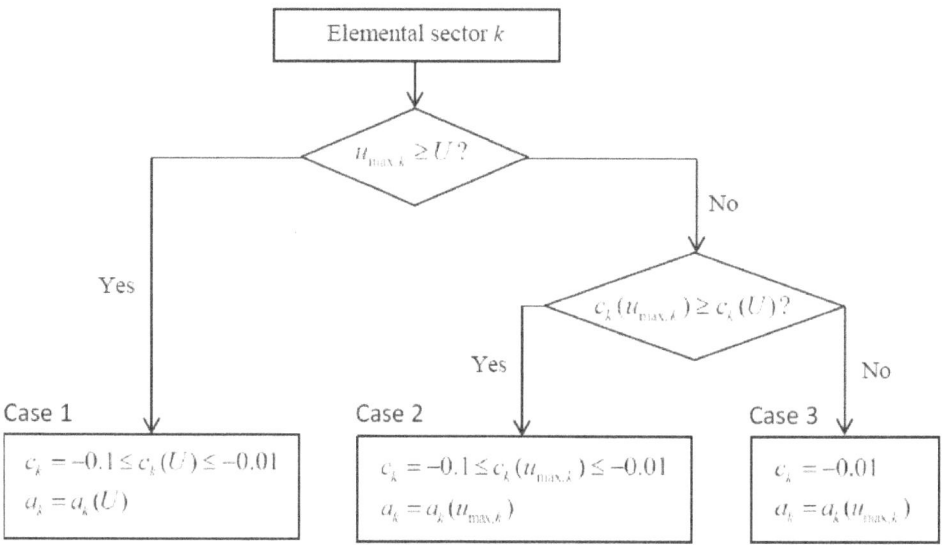

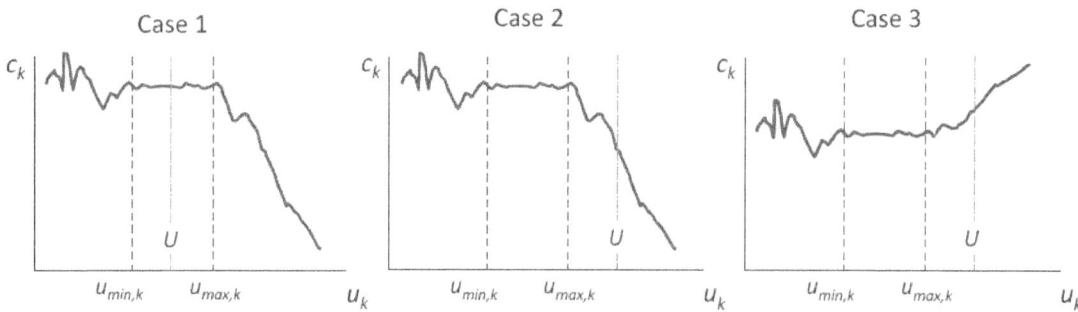

Fig. 2 Procedure for estimation of GPD parameters

3. Application

3.1 *Small Sets of Measured Directional Wind Speed Data*

We employ the proposed probabilistic model of directional wind speeds to generate synthetic synoptic wind speed series data for large MRIs at Newark, New Jersey. For the calibration of the proposed model for directional synoptic wind speeds, we use observed wind speed data in 36 directions in 10° increments, taken from the Automated Surface Observing System (ASOS), a network of about 20000 standardized US weather stations (NCDC, 2011). The ASOS synoptic wind data are adjusted by transforming knots into mph, using wind speeds for 18 elemental sectors with 20° increments, at 10 m above the ground (Fig. 3). The data have a threshold wind speed of approximately 29 mph (i.e., ≈ 25 knots), and a length of record of 35.34 years (from January 1, 1977 to April 24, 2012). Thus, the annual rate of occurrence for the wind events is 1212/35.34 year = 34.295 year^{-1}. Figure 4 shows the distribution of directional speeds of synoptic winds at Newark, NJ.

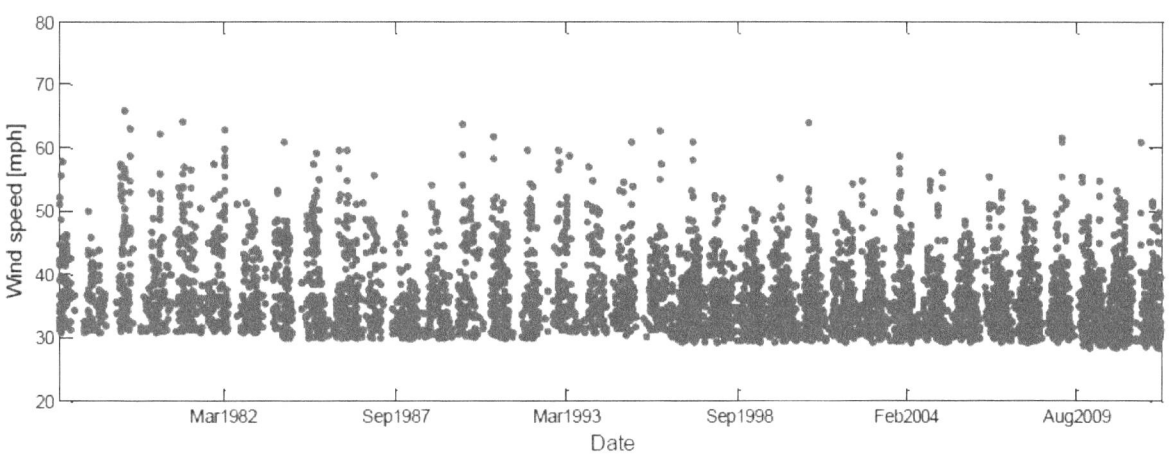

Fig. 3 Series of synoptic wind speeds, ASOS data (Newark, NJ)

This study uses a minimum number of 200 wind events for each directional sector. Because the directional data in each elemental sector are not sufficient to estimate parameters of the generalized Pareto distribution, some elemental sectors with too few data for statistical analysis are consolidated (i.e., 10° to 50°, 50° to 210°, and 210° to 250°) so that each consolidated sector has a sufficient number of data. Figure 4 shows the histogram of the directional wind speeds of the wind data and directional sectors determined by the requirement of minimum 200 windstorms. The vertical bars represent the number of wind speeds in each elemental sector. The horizontal line indicates the threshold number (200) of windstorm events. If the number of windstorm events in an elemental sector is lower than the threshold, the sector should be consolidated with one or more adjacent sectors to obtain a number of windstorm events with speeds higher than the threshold. The figure also shows that the data have 9 directional sectors that consist of 6 elemental sectors and 3 consolidated sectors. The histogram of the wind speeds for each directional sector is shown in Fig. 5.

7

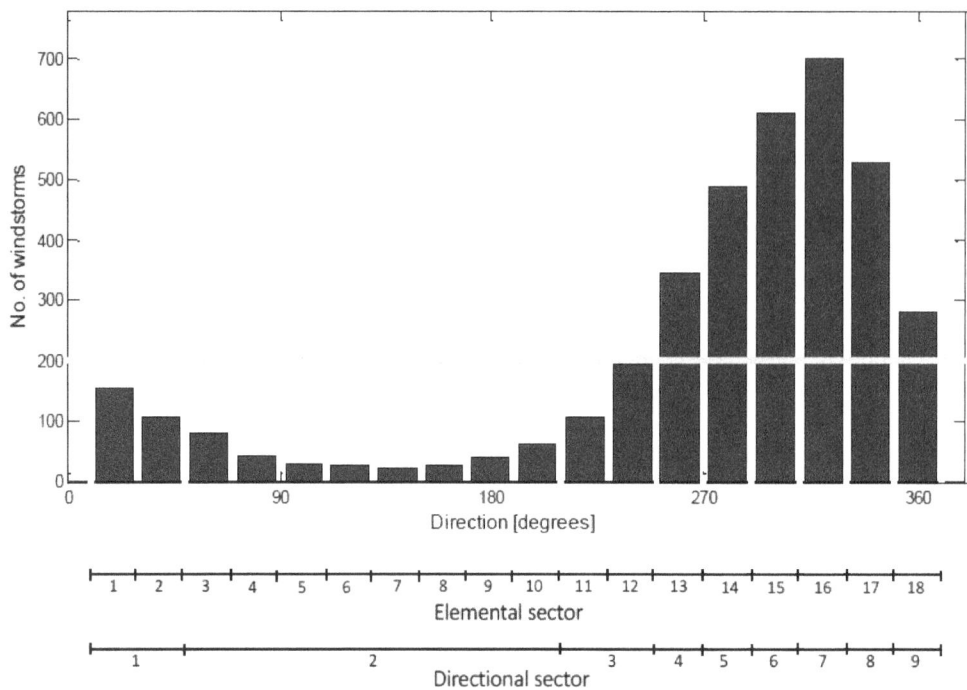

Fig. 4 Histogram of measured wind speeds of synoptic wind data (Newark, NJ)

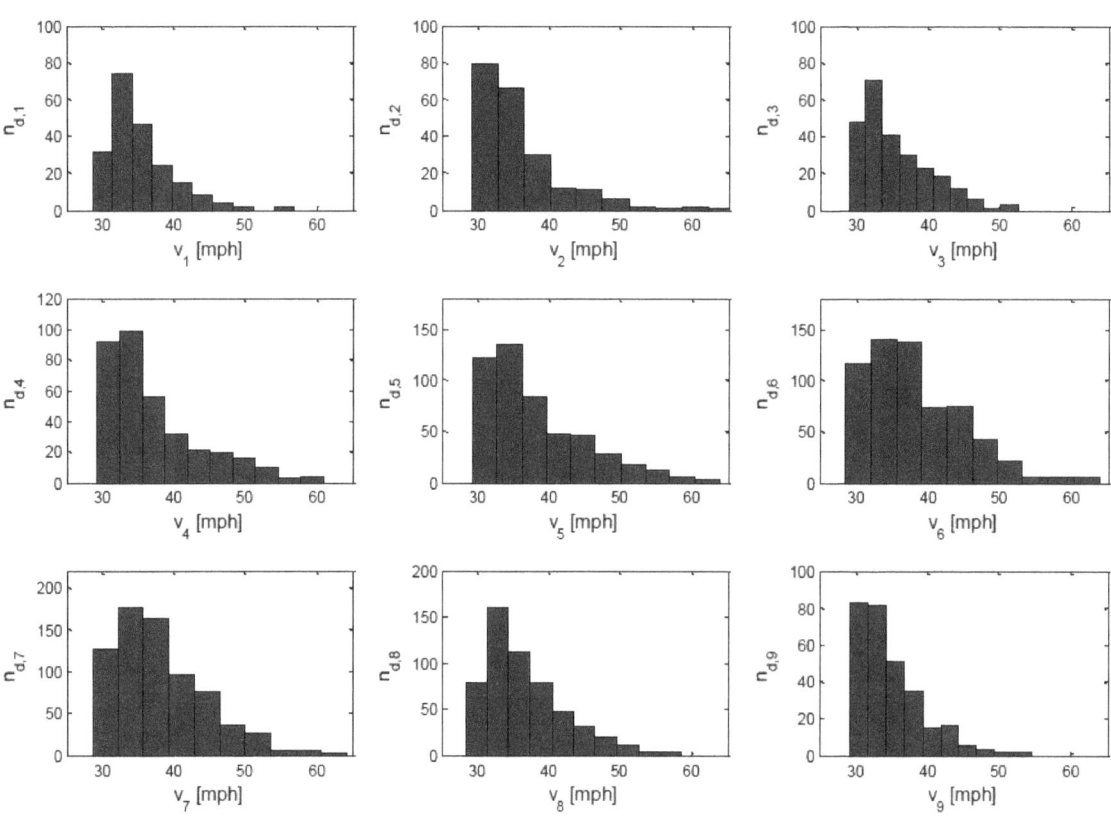

Fig. 5 Histogram of measured wind speeds in directional sectors

3.2 GPD Parameter Estimation

For the directional sector i, the GPD parameters (c_i, a_i,) based on the threshold wind speed u_i are estimated by the de Haan method (de Haan 1994) and are summarized in Table 1. The range of threshold wind speeds from $u_{min,i}$ to $u_{max,i}$ for constant c_i is obtained from the plots of GPD parameters based on the respective threshold wind speeds (Fig 6). The maximum of the $u_{min,i}$ values is $U = 37$ mph.

The tail length parameter c_i for directional sector i is calculated from the original c_i as a function of $u_{max,i}$ and U, as explained in Sect. 2.4. For the first directional sector, because $u_{max,1} = 40$ mph is higher than $U = 37$ mph, c_1 becomes $c_1(u_{max,1}) = -0.003$, but as noted in Sect. 2.2, this value is adjusted to -0.01; the associated scale parameter a_1 becomes $a(u_{max,1}) = 4.051$ mph.

The number of windstorm events whose non-directional highest speed is larger than U is 644, and the associated mean arrival rate λ is 18.22 (= 644 storms/35.34 years). For directional sector i the number of wind speeds above U, n_i, is calculated using the expression $n_i = (\lambda_i/\lambda)r$, where the ratio of (λ_i/λ) is the probability of wind speeds for sector i being higher than the threshold U, (i.e., $q_i = P(v_i > U)$). Once the GPD parameters are determined, the wind speeds corresponding to MRIs can be obtained using Eq.(2) as shown in Fig. 8. Figure 8 shows, for comparison, wind speeds obtained by using: 1) original parameters, 2) adjusted parameters as explained above, and 3) $c = -0.01$ and the original a that results in wind speeds based on Gumbel-like distribution. For all directional sectors except the first one, the wind speeds estimated from the adjusted parameters are higher than the wind speeds from the original parameters and lower than the wind speeds based on the Gumbel parameters. These results are consistent with the purpose of the method that uses adjusted parameters, which yields reasonably conservative wind speeds for design of structures.

Table 1. Estimated parameters of generalized Pareto distributions

Directional sector	$u_{min,i}$ [mph]	$u_{max,i}$ [mph]	Original c_i	Original a_i [mph]	Adjusted c_i	Adjusted a_i [mph]	λ_i [year^{-1}]
1	34	40	-0.003	4.051	-0.010	4.051	1.585
2	33	39	-0.042	6.002	-0.042	6.002	1.641
3	33	41	-0.305	5.321	-0.100	5.321	2.235
4	37	49	-0.352	8.879	-0.100	8.879	3.650
5	37	52	-0.273	8.291	-0.100	8.291	6.112
6	35	46	-0.249	7.186	-0.100	7.186	8.546
7	37	47	-0.163	6.196	-0.100	6.196	10.045
8	34	44	-0.201	5.782	-0.100	5.782	5.914
9	33	42	-0.135	4.596	-0.100	4.596	2.037

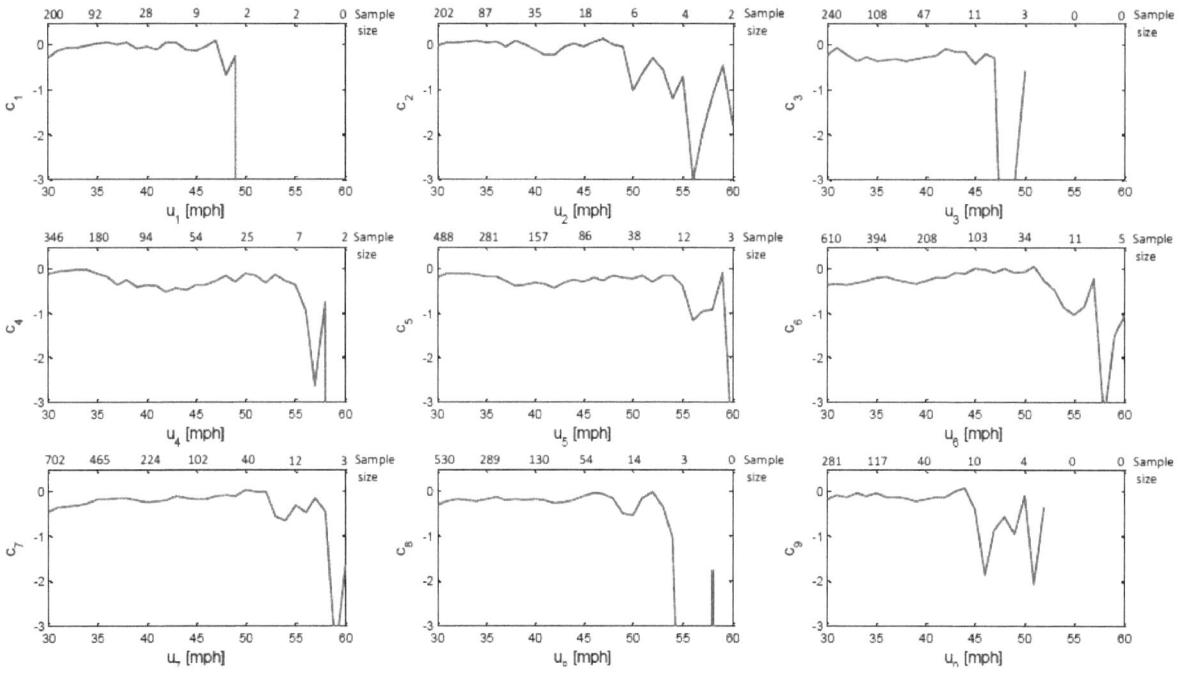

Fig. 6 Estimated tail length parameters versus threshold and number of threshold exceedances

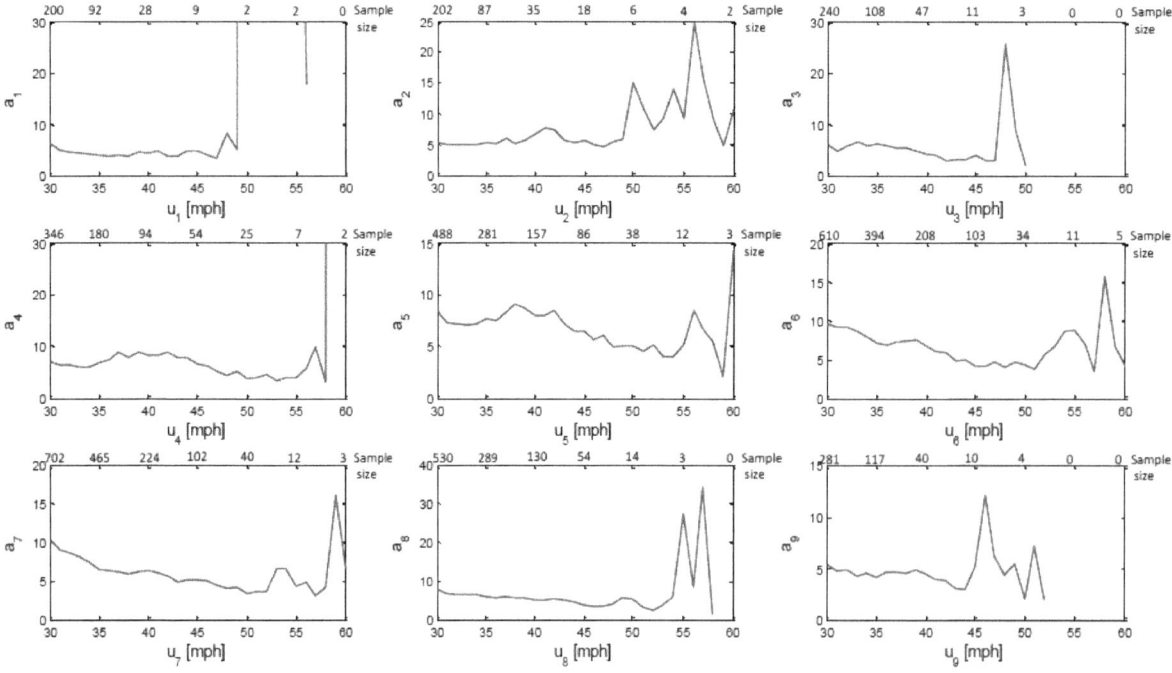

Fig. 7 Estimated scale parameters versus threshold and number of threshold exceedances

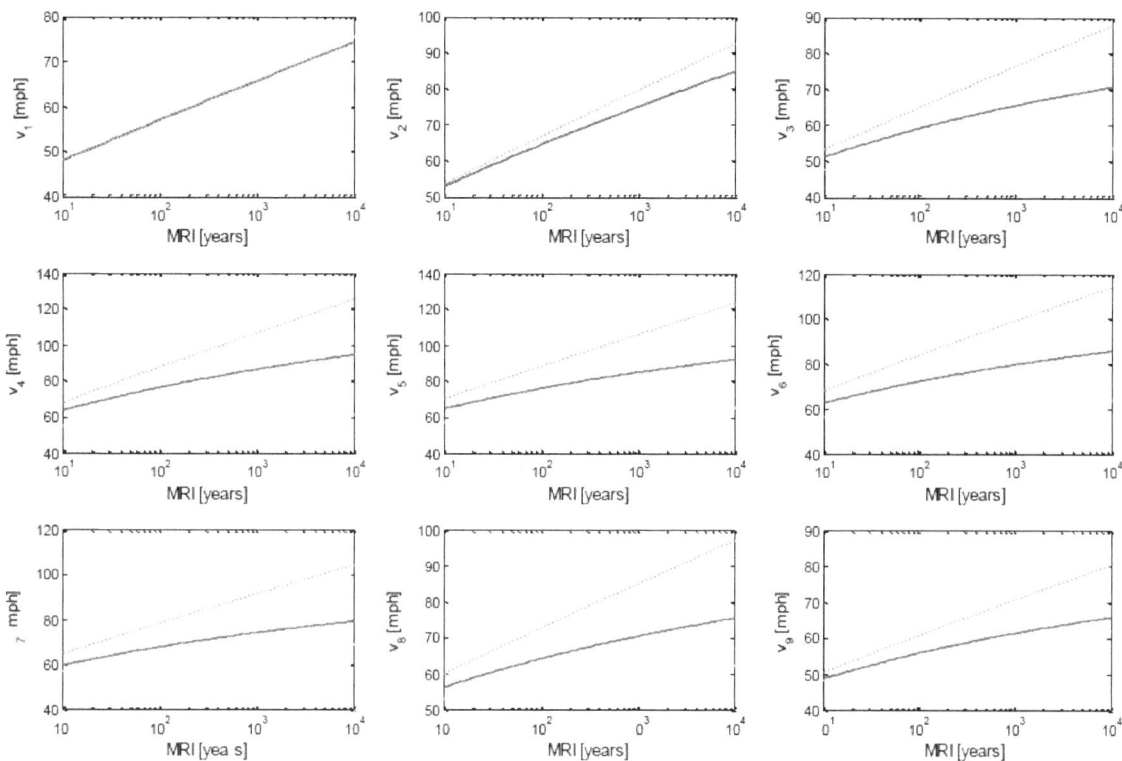

Fig. 8 Wind speeds in directional sectors as functions of MRIs
(for original c;   for adjusted c;   for $c = -0.01$)

3.3 Generation of Large Directional Wind Speed Datasets by Monte Carlo Simulation

Synthetic wind speeds for synoptic wind events in Newark, NJ have been generated in all elemental sectors by Monte Carlo simulation using adjusted directional sector parameters estimated by the de Haan method, as shown in the previous section. This study generates the time-series of directional synoptic wind speeds for 10,000 years to assure a precise estimation of the wind speed and the associated wind effects with MRIs of up to 1,700 years.

The synthetic data for synoptic wind events covering a period of 10,000 years correspond to $r = 211,900$ wind events (Eq.(3)) in 18 elemental directions. Assume that the elemental sector k belongs to the directional sector i. For elemental sector k the requisite number n_k of wind speeds above the threshold U is calculated by using $n_k = (\lambda_i/\lambda) \ r = r \ q_i$. The wind speeds above U are generated n_k times in by Monte Carlo simulation using Eq.(4) with estimated GPD parameters (c_i, a_i) in directional sector i. This approach provides conservative results for the elemental sectors that are associated with consolidated sectors. As shown in Figs. 9 and 10, for elemental sectors 1 to 12 belonging to the consolidated sectors 1, 2 and 3 (Fig. 4) the number of synthetic wind speeds is augmented. Figure 11 shows the distribution of the synthetic wind speeds in elemental sectors.

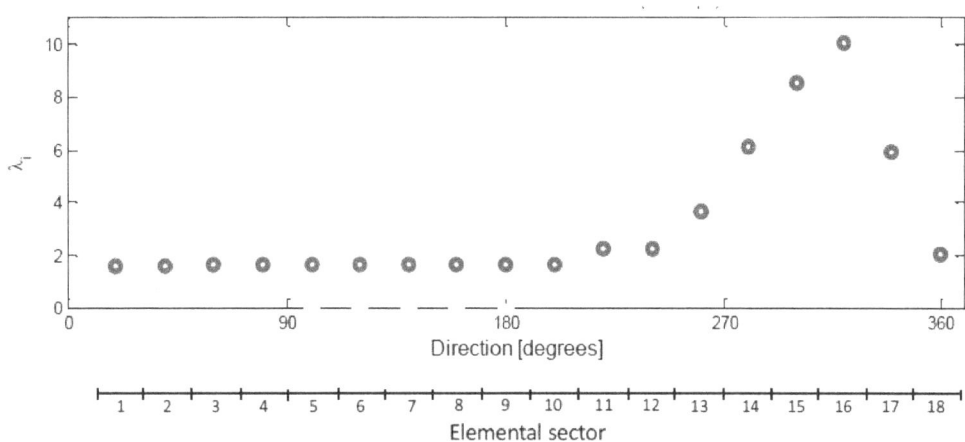

Fig. 9 Comparison of λ_i's in elemental sectors for measured data and synthetic data

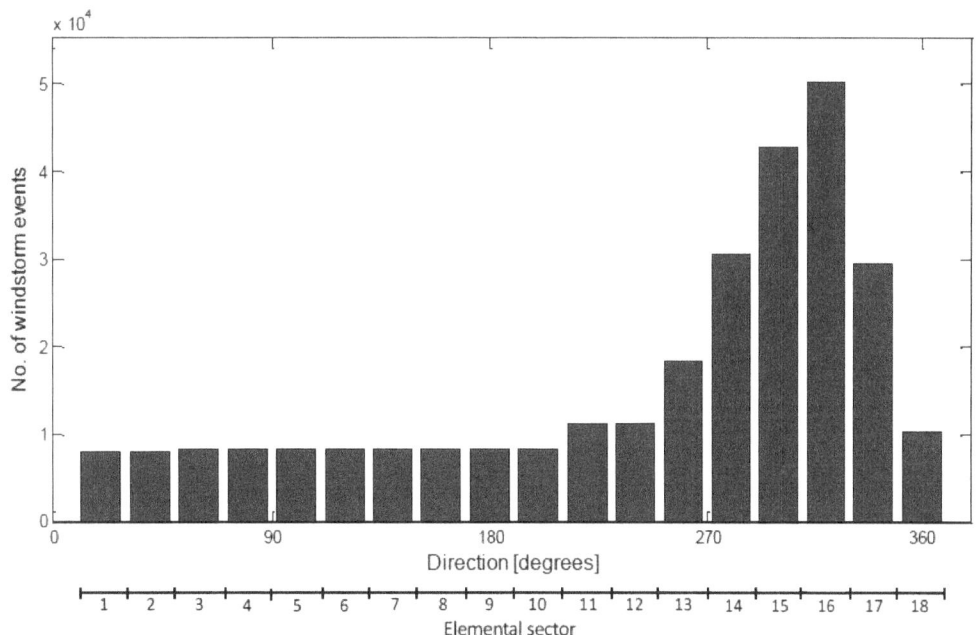

Fig. 10 Histogram of synthetic wind speed data for MRI = 10.000 years (Newark, NJ)

13

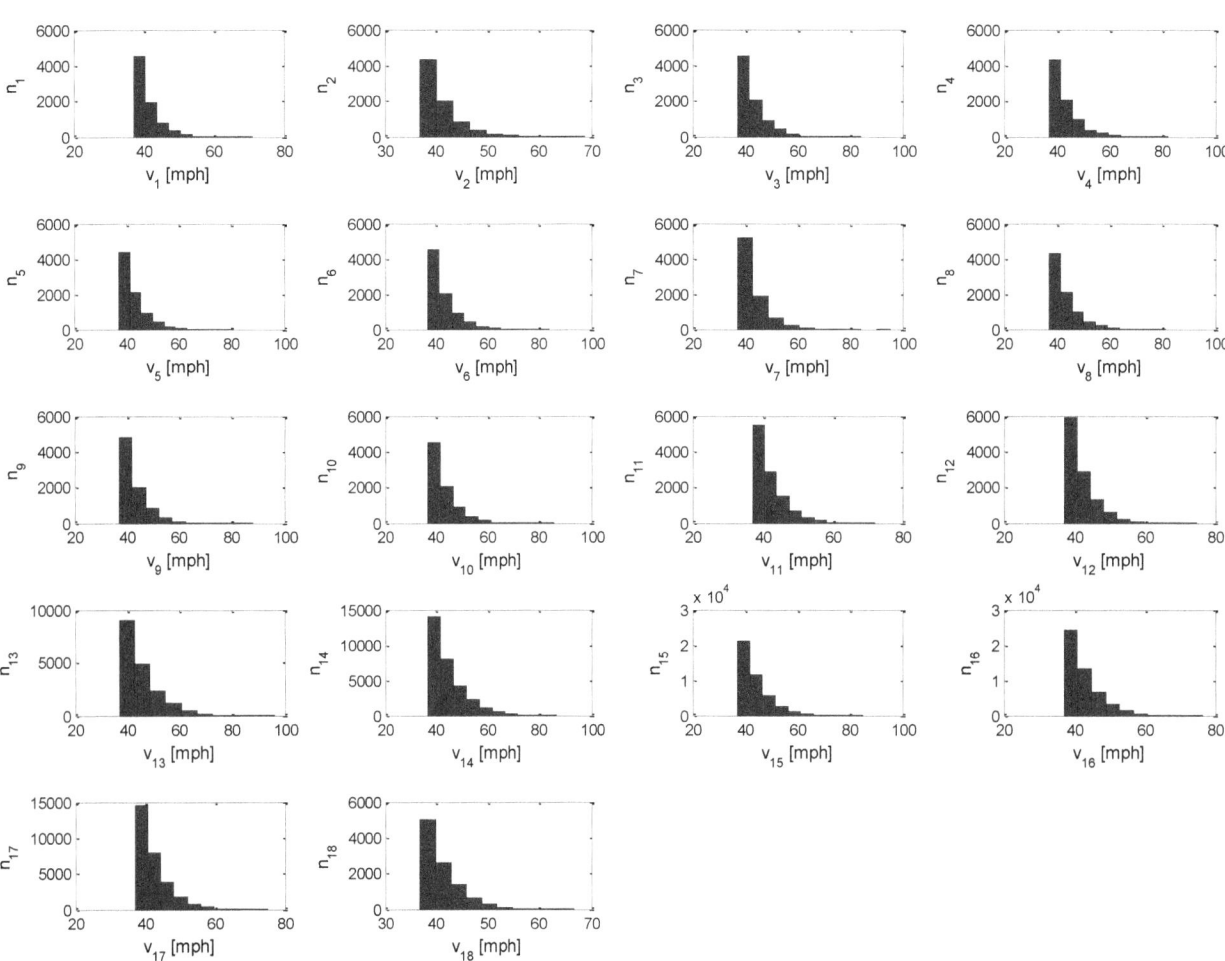

Fig. 11 Synthetic wind speeds in elemental sectors

4. Conclusions

We proposed an algorithm to generate large sets of directional synoptic or thunderstorm wind speeds of sufficient size to allow the estimation of wind effects with long mean recurrence intervals. The probability model in the study was based on a generalized Pareto distribution under the assumption of directionally independent wind speeds. The Monte Carlo simulation procedure that uses this probability model was described in detail. The algorithm applies to measured data sets that are too small to allow reliable estimates of correlations between pairs of sectorial directional wind speeds. Reference is made to earlier published results according to which the assumption of independence is marginally conservative provided that the center angles of the directional sectors being considered are sufficiently large (say, 20°), and reference is made to results applicable to directional data sets exhibiting large correlations (e.g., larger than 0.7). An application was presented, for which the parameters of the distribution were estimated from the ASOS data by the de Haan method. The proposed algorithm uses adjusted distribution parameters to prevent unconservative wind speed estimates. Future research includes the use of simulated data to estimate sampling errors in the estimation of wind-induced effects on structures as functions of the size of the measured data sample. The software required to perform the simulation is available on www.nist.gov/wind.

References

ASCE (2010). *Minimum design loads for buildings and other structures*, American Society of Civil Engineers, Reston, VA.

de Haan, L. (1994). "Extreme Value Statistics." in *Extreme value theory and applications*, J. Galambos, J. Lechner, and E. Simiu, eds., Kluwer Academic Publishers, 93-122.

Galambos, J., Lechner, J., and Simiu, E. (1994). "Extreme Value Theory and Applications.*" Kluwer Academic Publishers.

Grigoriu, M. (2009). Algorithms for generating large sets of synthetic directional wind speed data for hurricane, thunderstorm, and synoptic winds. NIST Technical Note 1626, National Institute of Standards and Technology, Gaithersburg, MD.

Lombardo, F. T., Main, J. A., and Simiu, E. (2009). "Automated extraction and classification of thunderstorm and non-thunderstorm wind data for extreme-value analysis." *Journal of Wind Engineering and Industrial Aerodynamics*, 97(3-4), 120-131.

National Climatic Data Center (NCDC) (2011). *Data documentation for data set 3505 (DSI-3505)*: <http://www1.ncdc.noaa.gov/pub/data/documentlibrary/tddoc/td3505.pdf> (accessed 05/17/12).

Simiu, E. (2011). *Design of buildings for wind: a guide for ASCE 7-10 Standard users and designers of special structures*, 2nd ed., John Wiley & Sons, Hoboken, NJ.

Simiu, E. and Scanlan, R. H. (1996). *Wind effects on structures*, 3rd ed., John Wiley & Sons.

Yeo, D. and Simiu, E. (2011). "High-rise reinforced concrete structures: Database-Assisted Design for wind." *Journal of Structural Engineering*, 137(11), 1340-1349.

www.ingramcontent.com/pod-product-compliance
Lightning Source LLC
Chambersburg PA
CBHW081821170526
45167CB00008B/3498